HRH Prince Joe Duncan Wright

Digitisation of steel manufacturing

1st Published in 2012

©Copyright

Table of Contents

Digitisation of steel manufacturing 3

Smart manufacturing 5

Automation and robotics 10

Data Collection, connectivity, data analytics and AI **16**

Digital Twin 34

Supply chain optimisation 40

Energy efficiency and sustainability **50**

Continuous improvement and optimisation 57

Quality control and inspection 63

Collaboration and integration **69**

Advantages of digitisation 81

The future of steel manufacturing 88

Digitisation of steel manufacturing

The digitisation of steel manufacturing, often referred to as digital transformation or Industry 4.0 in the steel industry, involves the integration of digital technologies and data-driven systems to optimize and enhance various aspects of the steel manufacturing process. It encompasses the use of advanced sensors, automation, data analytics, artificial intelligence (AI), and other digital tools to improve efficiency, productivity, quality, and sustainability in steel production.

The digitisation of steel manufacturing, also known as digital transformation or Industry 4.0 in the steel industry, involves the integration of digital technologies and data-driven processes to optimize and enhance various aspects of steel production. It aims to improve efficiency, productivity, quality, sustainability, and safety within steel manufacturing operations.

The digitisation of steel manufacturing refers to the integration of digital technologies and data-driven processes in the steel production and manufacturing industry. It involves the use of advanced technologies such as artificial intelligence (AI), Internet of Things (IoT), big data analytics, robotics, and automation to optimize and streamline various stages of the steel manufacturing process.

Smart Manufacturing

Digitisation enables the implementation of smart manufacturing practices, where sensors and internet-connected devices gather real-time data from various stages of the steel production process. This data is analysed using advanced analytics and AI algorithms to optimize production, monitor equipment performance, predict maintenance needs, and identify process improvements.

Internet of Things (IoT) and Sensor Integration is a key aspect of smart manufacturing in the digitisation of steel manufacturing. IoT sensors are deployed throughout the steel manufacturing facility to collect real-time data from equipment, machinery, and processes. These sensors monitor parameters such as temperature, pressure, humidity, vibration, and energy consumption. The data is transmitted to a central system for analysis and decision-making.

Data Analytics and Artificial Intelligence (AI) is another key aspect of smart manufacturing in the digitisation of steel manufacturing. The collected data is analysed using advanced analytics and AI techniques. Machine learning algorithms can identify patterns, anomalies, and correlations within the data to optimize processes, improve efficiency, and reduce downtime. Predictive analytics can also help identify potential equipment failures and recommend maintenance actions.

Real-Time Monitoring and Control is key in smart manufacturing in the digitisation of steel manufacturing: With smart manufacturing, steel manufacturers have real-time visibility into the entire production process. They can monitor and control parameters such as temperature, flow rates, and quality metrics in real-time. This enables quick intervention, adjustment, and optimisation of processes to ensure consistent quality and minimize waste.

Digital twin technology creates a virtual replica or simulation of the steel manufacturing process. By integrating real-time data and process models, manufacturers can simulate different scenarios and predict the impact of changes or improvements. This allows for optimisation of process parameters, product design, and resource allocation.

Smart manufacturing in steel production involves the integration of collaborative robotics and automation systems. Robots and automated systems can handle tasks such as material handling, assembly, welding, and quality inspection. They work alongside human workers, enhancing productivity, improving safety, and reducing errors.

Smart manufacturing integrates the steel manufacturing facility with the broader supply chain integration. Through digital connectivity and information sharing, manufacturers can achieve end-to-end visibility and coordination. This enables better demand forecasting, optimized inventory management, and seamless integration with suppliers and customers.

As digitisation and connectivity increase, ensuring the security of data and systems becomes crucial. Smart manufacturing in steel production incorporates robust cybersecurity measures to protect against cyber threats and ensure the integrity and confidentiality of sensitive data.

The adoption of smart manufacturing in steel production brings numerous benefits, including improved productivity, enhanced quality control, reduced downtime, optimized resource utilization, and increased flexibility. It empowers manufacturers to respond quickly to market demands, optimize processes, and make data-driven decisions for continuous improvement.

Automation and Robotics

Automation plays a significant role in digitizing steel manufacturing.

Robotic systems can perform tasks such as material handling, assembly, and quality control with precision and efficiency. Automated guided vehicles (AGVs) are used for transporting materials within the production facility, while robotic arms handle tasks like welding, cutting, and painting.

Automation reduces human error, improves safety, and increases production throughput. Digitisation enables the automation of processes in steel manufacturing, reducing manual labour and increasing efficiency. Robotics and autonomous systems can be employed for tasks such as material handling, welding, inspection, and maintenance. Automated guided vehicles (AGVs) and robots can work alongside human workers to improve safety and productivity.

Automation and robotics play a significant role in the digitisation of steel manufacturing, offering several advantages in terms of efficiency, precision, and safety. Material Handling is a key aspect in automation and robotics. Automation systems, such as robotic arms and automated guided vehicles (AGVs), are used to handle heavy raw materials and finished products. These systems can transport materials within the facility, load and unload trucks, and feed materials into various stages of the manufacturing process. They improve efficiency, reduce manual labour, and enhance workplace safety.

Robotic welding and cutting systems are widely employed in steel manufacturing. These robots can perform complex welding operations with high precision and consistency, ensuring strong and accurate welds. Automated cutting systems, such as laser cutting machines, enable precise and efficient material cutting, reducing waste and improving productivity.

Assembly and manipulation is another key aspect in automation and robotics. Robotic systems are used for the assembly and manipulation of steel components. They can handle intricate assembly tasks, such as joining complex structures or fastening components together. Robots offer speed, accuracy, and repeatability, ensuring consistent quality in assembly processes.

Robotics and automation systems are utilized for inspection and quality inspection in steel manufacturing. Vision systems and sensors integrated with robots can perform non-destructive testing, measure dimensions, detect defects, and ensure adherence to quality standards. Automated inspection processes reduce human error and enhance the reliability and consistency of quality control.

Automation systems are employed for machine tending tasks, such as loading and unloading materials into machines and equipment. Robots can operate alongside machinery, ensuring uninterrupted production, reducing cycle times, and freeing up human operators for more complex tasks.

Robotics and automation systems are integrated with data collection and analysis processes.

Sensors and monitoring devices on robots collect real-time data on process parameters, equipment performance, and product quality. This data is analysed using advanced analytics techniques to identify patterns, optimize processes, and improve overall manufacturing efficiency. Automation and robotics contribute to improved workplace safety and ergonomics in steel manufacturing. By taking over repetitive, physically demanding, or hazardous tasks, robots reduce the risk of injuries for human workers. They can operate in harsh environments, such as extreme temperatures or toxic atmospheres, ensuring worker safety and well-being.

By integrating automation and robotics into steel manufacturing processes, digitisation enhances productivity, reduces costs, improves product quality, and enhances workplace safety. These technclogies enable manufacturers to optimize operations, respond to changing market demands, and achieve higher levels of efficiency and competitiveness.

Data Collection, Connectivity, Data Analytics and AI

The vast amount of data generated in steel manufacturing can be leveraged through data analytics and AI techniques. By analysing historical and real-time data, AI algorithms can identify patterns, optimize process parameters, and make predictions. This leads to better quality control, improved resource allocation, reduced energy consumption, and enhanced overall process efficiency. Digitisation involves collecting data from various sources within the steel manufacturing process, such as sensors, machines, and production systems. This data is then connected and transmitted through the Internet of Things (IoT) and other communication networks to create a digital infrastructure.

Digitisation enables the use of advanced analytics and machine learning algorithms to analyse the collected data and gain insights. Predictive maintenance techniques can be implemented to detect potential equipment failures, optimize maintenance schedules, and reduce downtime. Data analytics and artificial intelligence (AI) play a crucial role in the digitisation of steel manufacturing, enabling advanced insights, predictive capabilities, and optimisation of various processes.

Data from various sources, such as sensors, production equipment, and enterprise systems, are collected and integrated into a centralized data repository. This includes data on process parameters, equipment performance, energy consumption, quality metrics, and supply chain information. In the digitisation of steel manufacturing, data from various sources, including sensors, production equipment, and enterprise systems, is collected and integrated into a centralized data repository. This data repository serves as a central hub where all the collected data is stored, organized, and made accessible for analysis and decision-making purposes.

Sensors installed throughout the steel manufacturing facility collect real-time data on various parameters such as temperature, pressure, vibration, humidity, and energy consumption. These sensors can be embedded in equipment, machinery, production lines, or even wearables for monitoring worker conditions.

Production equipment, such as furnaces, conveyors, and robots, often have built-in data collection capabilities. They generate data on operational performance, machine status, production rates, and other relevant metrics. This data is captured and transmitted to the centralized repository. Data from enterprise systems, such as Enterprise Resource Planning (ERP) systems or Manufacturing Execution Systems (MES), is integrated into the centralized data repository. This includes data on inventory, from both the shop floor and the business operations are consolidated in one place.

Data is transmitted from sensors, equipment, and systems to the centralized repository using various communication protocols such as Ethernet, Wi-Fi, or industrial automation protocols like OPC (OLE for Process Control) or MQTT (Message Queuing Telemetry Transport). This ensures a seamless flow of data from multiple sources. Once the data is collected, it may undergo transformations and preprocessing to ensure consistency and compatibility. This may involve cleaning the data, normalizing units of measurement, or converting formats. The transformed data is then stored in the centralized data repository, which can be a data warehouse, a cloud-based storage system, or a dedicated server.

Data governance policies and security measures are implemented to ensure data integrity, privacy, and compliance. Access controls and authentication mechanisms are established to safeguard the data repository from unauthorized access. Data governance practices, such as data quality management and data lifecycle management, are employed to maintain the accuracy, completeness, and relevance of the data. By integrating data from sensors, production equipment, and enterprise systems into a centralized data repository, steel manufacturers can gain a holistic view of their operations. This integrated data serves as a foundation for data analysis, machine learning, and AI techniques, enabling manufacturers to extract valuable insights, optimize processes, and make informed decisions to drive efficiency and productivity in steel manufacturing.

Advanced data analytics techniques are applied to the collected data to derive meaningful insights. Descriptive analytics techniques help visualize and summarize historical data, providing a clear understanding of past performance and trends. Visualization tools present data in a user-friendly format, facilitating easy interpretation and decision-making.

Predictive analytics uses statistical models and machine learning algorithms to analyse historical data and predict future outcomes. In steel manufacturing, predictive analytics can forecast equipment failures, maintenance needs, quality issues, and production bottlenecks.

This enables proactive decision-making, reduces downtime, and enhances overall operational efficiency. AI and data analytics are employed to optimize various processes in steel manufacturing. By analysing process data, patterns, and correlations, manufacturers can identify areas for improvement, reduce energy consumption, optimize resource allocation, and minimize waste. Optimized processes lead to increased productivity, cost savings, and improved product quality.

Data analytics and AI techniques are used to monitor and control product quality in real-time. Machine learning algorithms can analyse sensor data to detect anomalies, identify potential quality issues, and trigger immediate corrective actions. This ensures consistent product quality and reduces the likelihood of defects.

Data analytics and artificial intelligence (AI) are instrumental in the digitisation of steel manufacturing, providing valuable insights, predictive capabilities, and optimisation opportunities. Data from various sources, such as sensors, equipment, and enterprise systems, is collected and integrated into a centralized repository. This data includes information on process parameters, equipment performance, quality metrics, energy consumption, and supply chain data.

Descriptive analytics involves analysing historical data to understand past performance and trends. It helps in visualizing and summarizing data, enabling manufacturers to gain insights into production patterns, identify areas of improvement, and understand factors influencing productivity and quality.

Predictive analytics plays a vital role in the digitisation of steel manufacturing by leveraging historical data and statistical models to make predictions about future outcomes. It enables manufacturers to anticipate and address potential issues, optimize processes, and improve overall operational efficiency. Predictive analytics can help optimize maintenance activities by forecasting equipment failures. By analysing historical data on equipment performance, sensor readings, and maintenance records, predictive models can identify patterns and indicators of potential failures. This enables manufacturers to schedule maintenance proactively, reducing unplanned downtime and optimizing maintenance resources. Predictive analytics can assist in optimizing production planning and scheduling processes. By analysing historical production data, market demand patterns, and supply chain factors, manufacturers can predict future demand and optimize production schedules accordingly. This helps in achieving efficient resource utilization, minimizing production bottlenecks, and meeting customer demands in a timely manner.

Predictive analytics can aid in quality control by identifying potential quality issues before they occur. By analysing historical quality data, process parameters, and sensor readings, manufacturers can develop models that predict the likelihood of quality defects or deviations. This enables early detection and intervention, reducing scrap, rework, and customer complaints. Predictive analytics can be used to optimize energy consumption in steel manufacturing. By analysing historical energy usage data, weather conditions, and production patterns, manufacturers can develop models to predict energy demand and identify opportunities for energy optimisation. This helps in reducing energy costs, improving sustainability, and achieving energy efficiency targets.

Predictive analytics can contribute to supply chain optimisation in steel manufacturing. By analysing historical data on demand patterns, inventory levels, and market conditions, predictive models can forecast future demand and optimize inventory levels accordingly. This helps in minimizing stockouts, reducing excess inventory, and optimizing supply chain logistics.

Predictive analytics can optimize various processes within steel manufacturing. By analysing historical process data, sensor readings, and environmental factors, predictive models can identify process inefficiencies, bottlenecks, or deviations from optimal conditions. Manufacturers can then make data-driven decisions to optimize process parameters, reduce waste, and improve overall productivity. Predictive analytics can aid in safety and risk management within steel manufacturing.

By analysing historical safety data, incident reports, and environmental conditions, predictive models can identify potential safety risks and forecast the likelihood of accidents. This enables manufacturers to implement preventive measures, enhance safety protocols, and reduce workplace incidents.

By leveraging predictive analytics, steel manufacturers can proactively address challenges, optimize operations, and improve overall performance. It empowers manufacturers to make informed decisions based on data-driven insights, leading to increased efficiency, reduced costs, enhanced product quality, and improved safety in the digitized steel manufacturing environment.

Predictive analytics leverages statistical models and machine learning algorithms to analyse historical data and make predictions about future outcomes. In steel manufacturing, predictive analytics can be used to forecast equipment failures, optimize maintenance schedules, predict production bottlenecks, and anticipate quality issues. This enables proactive decision-making and minimizes downtime.

Data analytics and AI techniques are employed to optimize various processes in steel manufacturing. By analysing data, identifying patterns, and applying optimisation algorithms, manufacturers can optimize process parameters, minimize waste, reduce energy consumption, and improve overall efficiency. This leads to increased productivity, cost savings, and enhanced product quality. Data analytics and AI enable real-time monitoring of quality parameters during steel manufacturing processes. Machine learning algorithms can analyse sensor data in real-time, detecting anomalies, identifying potential quality issues, and triggering immediate corrective actions. This ensures consistent product quality, minimizes defects, and reduces scrap or rework.

Data analytics and AI techniques help optimize the steel manufacturing supply chain. By analysing data on demand patterns, inventory levels, transportation logistics, and market conditions, manufacturers can optimize inventory management, production scheduling, and distribution. This leads to improved supply chain efficiency, reduced costs, and enhanced customer satisfaction.

Prescriptive analytics takes data analytics and AI to the next level by providing recommendations for optimal actions. By considering multiple variables, constraints, and business objectives, prescriptive analytics suggests the best course of action to achieve desired outcomes.

In steel manufacturing, prescriptive analytics can assist in optimizing production plans, resource allocation, maintenance schedules, and supply chain decisions. By leveraging data analytics and AI, steel manufacturers can unlock valuable insights, optimize processes, improve product quality, and make data-driven decisions. The digitisation of steel manufacturing enables the collection and analysis of data in real-time, empowering manufacturers to continuously improve operations, increase efficiency, and stay competitive in the industry.

Data analytics and AI help optimize the steel manufacturing supply chain. By analysing data on demand patterns, inventory levels, transportation logistics, and market conditions, manufacturers can make informed decisions regarding inventory management, production scheduling, and distribution. This minimizes stockouts, reduces lead times, and improves overall supply chain efficiency.

Prescriptive analytics takes data analysis and AI a step further by providing recommendations for optimal actions. By considering multiple variables and constraints, prescriptive analytics can suggest the best course of action to achieve specific goals. In steel manufacturing, prescriptive analytics can assist in optimizing production plans, resource allocation, and maintenance schedules. By leveraging data analytics and AI, steel manufacturers can gain valuable insights, optimize processes, enhance product quality, reduce costs, and improve overall operational efficiency. Digitisation enables the continuous monitoring and analysis of data, enabling timely decision-making and driving continuous improvement in steel manufacturing.

Digital Twin

A digital twin is a virtual representation of a physical asset or system, such as a steel mill or production line. It allows real-time monitoring and simulation of the manufacturing process, enabling operators to visualize and optimize operations digitally. By using digital twins, operators can test and evaluate different scenarios, identify bottlenecks, optimize scheduling, and minimize downtime.

Digital twins are virtual representations of physical assets or processes. They enable real-time monitoring, analysis, and optimisation of steel manufacturing operations. By simulating the behaviour of equipment, processes, and materials, manufacturers can optimize production parameters, reduce defects, and test new ideas before implementation.

Digital Twin technology is a key component of the digitisation of steel manufacturing.

A Digital Twin is a virtual representation of a physical steel manufacturing system, encompassing the entire lifecycle from design and construction to operation and maintenance. Digital Twin technology allows engineers to create a virtual model of the steel manufacturing system during the design phase. This includes modelling the equipment, processes, and material flows. Engineers can simulate different scenarios, test design changes, and optimize the system before physical implementation. It helps identify potential issues, reduce design flaws, and improve overall efficiency.

Digital Twin enables real-time monitoring of the physical steel manufacturing system by integrating it with sensors, IoT devices, and data analytics. The virtual model is continuously updated with real-time data, providing insights into the system's performance, operational parameters, and condition. This helps in early detection of anomalies, predictive maintenance, and optimisation of operations. By continuously monitoring the physical system through sensors and feeding the data to the Digital Twin, manufacturers can predict and prevent equipment failures.

The Digital Twin can analyse historical data, perform predictive analytics, and generate alerts or maintenance recommendations. This approach optimizes maintenance schedules, reduces downtime, and extends equipment lifespan.

Digital Twin technology enables manufacturers to optimize the steel manufacturing process. By analysing real-time data from the physical system, the Digital Twin can identify bottlenecks, inefficiencies, and opportunities for improvement.

Manufacturers can simulate and evaluate different process scenarios, adjust parameters, and optimize energy consumption, material usage, and production throughput. Digital Twin facilitates training anc skill development for operators and maintenance personnel. The virtual model can be used to simulate different operating conditions, emergency scenarios, and maintenance procedures. This helps in training personnel, improving their skills, and ensuring their readiness to handle real-world situations effectively.

Digital Twin provides a platform for continuous improvement in steel manufacturing. By comparing the performance of the physical system with its virtual representation, manufacturers can identify gaps, assess the impact of proposed changes, and implement improvements iteratively. It enables data-driven decision-making, enhances operational efficiency, and drives innovation.

Digital Twin supports the entire lifecycle management of the steel manufacturing system. From initial design to decommissioning, the Digital Twin provides a comprehensive view of the system, including historical data, maintenance records, and performance analytics. This information helps manufacturers optimize asset utilization, plan for upgrades or replacements, and ensure compliance with regulations.

Digital Twin technology revolutionizes steel manufacturing by combining real-time data, analytics, and simulation capabilities. It offers insights, predictive capabilities, and optimisation opportunities that enable manufacturers to enhance efficiency, reduce costs, improve quality, and accelerate innovation in the digitized steel manufacturing environment.

Supply Chain Optimisation

Digitisation facilitates improved supply chain management in steel manufacturing. By integrating data from various stakeholders, including suppliers, manufacturers, and customers, it enables better demand forecasting, inventory management, and logistics planning. This results in reduced lead times, optimized material flow, and improved coordination throughout the supply chain. Digitisation allows for better coordination and optimisation of the steel supply chain. Real-time data sharing and visibility across the entire value chain facilitate efficient production planning, inventory management, and logistics.

Supply chain optimisation is a crucial aspect of digitisation in the steel manufacturing industry. By leveraging digital technologies and data-driven approaches, manufacturers can optimize their supply chain operations, enhance efficiency, reduce costs, and improve overall performance. Digital technologies enable manufacturers to collect and analyse data on customer demand patterns, market trends, and historical sales data. Advanced analytics techniques, such as predictive modelling and machine learning, can be applied to forecast future demand accurately. This helps manufacturers optimize production planning, inventory management, and procurement activities to meet customer demands while minimizing excess inventory and stockouts.

Demand forecasting is a critical component of digitisation in steel manufacturing. By leveraging digital technologies and advanced analytics, manufacturers can improve the accuracy of demand forecasts, enhance production planning, optimize inventory levels, and ensure customer satisfaction.: Digitisation enables the collection and integration of data from various sources, including historical sales data, customer orders, market trends, economic indicators, and external data such as weather patterns or construction activity. This diverse set of data is consolidated and processed to develop a comprehensive view of demand drivers.

Digital technologies and advanced analytics techniques, such as machine learning and predictive modelling, are applied to analyse the integrated data. These techniques uncover patterns, correlations, and seasonality in historical data, enabling the development of accurate demand forecasting models. Digitisation enables the capture and analysis of real-time data, such as customer orders, market dynamics, and competitor activities. This real-time data is continuously fed into the demand forecasting models, allowing for adjustments and adaptations based on the latest information.

Digital technologies facilitate the segmentation of customers based on their purchasing behaviour, preferences, and other relevant factors. By understanding the unique characteristics and demands of different customer segments, manufacturers can tailor their demand forecasts and production plans accordingly. Digitisation enables collaboration and information sharing among different stakeholders in the supply chain, including customers, suppliers, and internal teams. Collaborative planning platforms allow for the exchange of demand-related information, feedback, and insights, which can enhance the accuracy of demand forecasts and drive collaborative decision-making.

Digital tools and simulations enable manufacturers to perform sensitivity analysis on demand forecasts. By assessing the potential impact of different scenarios, such as changes in market conditions, pricing strategies, or customer behaviour, manufacturers can understand the risks and uncertainties associated with demand forecasting and develop contingency plans.

Accurate demand forecasts derived from digitisation can be used to optimize production planning. Manufacturers can align production schedules, capacity utilization, and raw material procurement based on the forecasted demand, minimizing inventory carrying costs, production bottlenecks, and stockouts. Digitisation enables real-time visibility into inventory levels and demand fluctuations. By integrating demand forecasts with inventory management systems, manufacturers can optimize inventory levels, reduce excess stock, and ensure sufficient stock availability to meet customer demand. Through digitisation, manufacturers can collect feedback and data on the accuracy of demand forecasts. This information can be analysed to refine and improve forecasting models over time, enhancing their accuracy and reliability.

By leveraging digitisation and advanced analytics for demand forecasting, steel manufacturers can optimize their production planning, inventory management, and customer service. Accurate demand forecasts enable manufacturers to align their operations with market demand, reduce costs, improve customer satisfaction, and gain a competitive edge in the industry.

Digitisation allows manufacturers to have real-time visibility and control over inventory levels throughout the supply chain. By integrating data from production systems, warehouses, and suppliers, manufacturers can optimize inventory levels, reduce carrying costs, and minimize the risk of stockouts. IoT sensors and RFID technology can be used to track and monitor inventory in real-time, enabling proactive inventory management.

Digitisation facilitates seamless collaboration and communication with suppliers. Through digital platforms and tools, manufacturers can share real-time information, collaborate on production plans, and track supplier performance. This improves transparency, reduces lead times, and enhances supplier relationships, leading to better supply chain coordination and efficiency.

Digitisation enables optimisation of logistics operations, including transportation and distribution. By integrating data from various sources such as GPS, telematics, and traffic information, manufacturers can optimize routing, vehicle utilization, and delivery schedules. Advanced analytics can help identify cost-effective transportation options, reduce fuel consumption, and improve overall logistics efficiency.

Digitisation provides real-time visibility into the movement of materials, components, and finished goods across the supply chain. Through the use of IoT sensors, RFID tags, and barcode scanning, manufacturers can track and trace products at every stage of the supply chain. This enhances traceability, improves product safety, and enables quick response to disruptions or recalls.

Digitisation generates vast amounts of data across the supply chain. By leveraging data analytics techniques, such as data mining, predictive analytics, and optimisation algorithms, manufacturers can extract valuable insights, identify inefficiencies, and optimize supply chain processes. This includes optimizing production scheduling, warehouse layouts, distribution networks, and procurement strategies.

Digitisation enables manufacturers to proactively identify and mitigate supply chain risks. By analyzing data on factors like supplier performance, market trends, and geopolitical events, manufacturers can assess and manage risks effectively. Digital tools, such as simulation models, can be used to evaluate the impact of potential disruptions and develop contingency plans for business continuity.

By embracing digitisation and leveraging advanced technologies, steel manufacturers can optimize their supply chain operations, enhance agility, reduce costs, and improve customer satisfaction. The digitisation of the supply chain in steel manufacturing enables data-driven decision-making, better collaboration, and proactive management of supply chain processes, ultimately leading to a competitive advantage in the industry.

Energy Efficiency and Sustainability

Digitisation contributes to energy efficiency and sustainability in steel manufacturing. Real-time data monitoring and analysis help identify energy-intensive areas and optimize energy usage. Digital systems enable better control of emissions, waste management, and environmental compliance. Additionally, digitisation enables predictive maintenance, reducing unplanned downtime and optimizing resource utilization. Digitisation can help optimize energy consumption and reduce environmental impact in steel manufacturing. Real-time monitoring and analysis of energy usage enable the identification of energy-saving opportunities and the implementation of sustainable practices.

Energy efficiency and sustainability are important considerations in the digitisation of steel manufacturing. By adopting digital technologies and implementing sustainable practices, steel manufacturers can reduce energy consumption, minimize environmental impact, and improve overall sustainability. Digitisation enables real-time monitoring of energy consumption across the steel manufacturing processes. By integrating energy monitoring systems with digital platforms, manufacturers can collect data on energy usage, identify energy-intensive areas, and analyse energy patterns. This data helps in optimizing energy consumption, identifying energy-saving opportunities, and implementing energy-efficient practices.

Digital technologies, such as IoT sensors and automation systems, play a crucial role in energy efficiency. Sensors can monitor and control energy-intensive equipment, ensuring they operate at optimal levels.

Automation systems can optimize process parameters, adjust energy usage based on real-time demand, and reduce energy wastage.

Predictive analytics, coupled with energy data, can be utilized to develop models that forecast energy demand and identify potential energy inefficiencies. By analysing historical data, weather patterns, production schedules, and other factors, manufacturers can optimize energy usage, plan for peak energy demand periods, and minimize energy costs.

Digitisation allows steel manufacturers to integrate renewable energy sources into their operations. By monitoring energy generation from solar panels, wind turbines, or other renewable sources, manufacturers can optimize the utilization of renewable energy and reduce reliance on traditional energy sources. Digital platforms can facilitate the seamless integration and management of renewable energy systems. Digitisation enables the creation of virtual power plants, where multiple energy sources, including renewables, are integrated and managed efficiently. Energy management systems leverage data analytics and real-time monitoring to optimize energy distribution, storage, and usage, ensuring energy efficiency and reducing environmental impact.

Digital technologies can help identify opportunities for waste heat recovery in steel manufacturing processes. By monitoring and analysing process data, manufacturers can identify sources of waste heat and implement heat recovery systems. Waste heat can be utilized for various purposes, such as heating water, generating electricity, or powering other industrial processes, improving energy efficiency and reducing carbon emissions. Digitisation enables the implementation of life cycle assessment (LCA) methodologies to evaluate the environmental impact of steel manufacturing processes. By considering the entire life cycle of products, from raw material extraction to end-of-life, manufacturers can identify areas for improvement and implement sustainable practices to minimize environmental impact.

Digitisation facilitates the monitoring and optimisation of the entire supply chain to enhance sustainability. By integrating sustainability criteria into supplier selection processes, manufacturers can choose suppliers who adhere to environmental standards. Digital platforms can track and analyse supply chain data, such as transportation emissions or material sourcing, to identify opportunities for improvement and reduce the carbon footprint. Digitisation enables accurate and transparent reporting of environmental data. Manufacturers can collect, analyse, and report data on energy consumption, greenhouse gas emissions, water usage, and other environmental indicators. This helps in compliance with environmental regulations, meeting sustainability goals, and building trust with stakeholders.

By embracing digitisation and sustainable practices, steel manufacturers can achieve significant improvements in energy efficiency and sustainability. Digitisation enables real-time monitoring, optimisation of energy usage, integration of renewable energy sources, and sustainable practices throughout the supply chain, leading to reduced environmental impact and a more sustainable steel

manufacturing industry.

Continuous Improvement and Optimisation

The availability of data and digital tools enables continuous improvement initiatives in steel manufacturing. By monitoring and analysing process data, manufacturers can identify inefficiencies, implement corrective measures, and continuously optimize their operations. This leads to improved productivity, reduced costs, and enhanced product quality.

Continuous improvement and optimisation are integral to the digitisation of steel manufacturing. By leveraging digital technologies and data-driven approaches, manufacturers can continuously analyse and optimize various aspects of their operations to improve efficiency, productivity, quality, and overall performance. Digitisation enables the collection of vast amounts of data from various sources within the steel manufacturing process, including sensors, equipment, production systems, and enterprise systems. This data is analysed using advanced analytics techniques to uncover insights, identify patterns, and gain a deeper understanding of the manufacturing operations.

Real-time monitoring and data analytics provide manufacturers with a comprehensive view of the performance of their steel manufacturing processes. Key performance indicators (KPIs) such as production output, equipment downtime, energy consumption, and quality metrics can be continuously monitored. Deviations from desired performance levels can be identified, allowing for timely interventions and process adjustments. Digitisation enables manufacturers to optimize individual processes within steel manufacturing, such as raw material handling, smelting, refining, and shaping. By analysing process data and applying optimisation algorithms, manufacturers can identify bottlenecks, inefficiencies, and opportunities for improvement. Process parameters can be adjusted, equipment settings optimized, and workflows streamlined to enhance efficiency and productivity.

Digitisation allows manufacturers to optimize their supply chain operations, including procurement, inventory management, and logistics. By integrating data from suppliers, warehouses, and transportation systems, manufacturers can optimize inventory levels, reduce lead times, and minimize costs. Advanced analytics techniques can be applied to identify opportunities for process streamlining and cost reduction throughout the supply chain. Digitisation enables real-time quality monitoring and control throughout the steel manufacturing process. By integrating quality data, including sensor measurements, inspection results, and customer feedback, manufacturers can identify quality issues, deviations, and defects in real-time. Predictive analytics can be used to detect patterns and early warning signs of potential defects, enabling proactive measures for defect prevention and quality improvement.

Digitisation allows for predictive maintenance practices in steel manufacturing. By continuously monitoring equipment performance, collecting data on parameters such as vibration, temperature, and energy consumption, manufacturers can apply predictive analytics techniques to detect potential equipment failures before they occur. This enables proactive maintenance interventions, minimizing unplanned downtime and optimizing equipment lifespan. Digital technologies facilitate the use of simulation and modelling techniques in steel manufacturing. Virtual models, such as digital twins, can be created to simulate different scenarios, evaluate process changes, and optimize production parameters. This helps manufacturers make informed decisions, test hypotheses, and assess the impact of potential changes before implementing them in the physical manufacturing environment.

Digitisation enables a culture of continuous learning and adaptation in steel manufacturing. By analysing data, gathering insights, and implementing process improvements, manufacturers can iterate and refine their operations. Lessons learned from data analysis can be shared across the organisation, driving continuous improvement efforts and fostering a culture of innovation. Digitisation promotes data-driven decision-making across all levels of the organisation. By analysing real-time and historical data, manufacturers can make informed decisions, backed by data and insights. This minimizes reliance on intuition and anecdotal evidence, leading to more effective decision-making and improved overall performance.

By embracing digitisation and leveraging data-driven approaches, steel manufacturers can achieve continuous improvement and optimisation in their operations. Digitisation enables real-time monitoring, process optimisation, predictive maintenance, and data-driven decision-making, driving efficiency, productivity, and competitiveness in the steel manufacturing industry.

Quality Control and Inspection

Digitisation can improve quality control and inspection processes in steel manufacturing. Advanced imaging technologies, non-destructive testing methods, and artificial intelligence can be employed to detect defects, measure dimensions, and ensure adherence to quality standards.

Quality control and inspection play a crucial role in the digitisation of steel manufacturing. By leveraging digital technologies, manufacturers can enhance their quality control processes, improve product quality, and ensure compliance with industry standards. Digitisation enables real-time monitoring of quality parameters throughout the steel manufacturing process. Sensors and IoT devices can collect data on factors such as temperature, pressure, vibration, and composition at various stages of production. This real-time data provides insights into process deviations and allows for immediate corrective actions to maintain product quality.

Digital technologies, such as machine vision systems and automated inspection equipment, can be utilized for quality inspection. These systems use cameras, sensors, and advanced algorithms to detect defects, measure dimensions, and analyse surface finishes with high precision. Automated inspection reduces human error, improves inspection speed, and ensures consistent and accurate quality assessment.

Digitisation enables the application of non-destructive testing techniques for quality control. NDT methods, such as ultrasonic testing, magnetic particle inspection, and radiographic testing, can be enhanced through digital technologies. Digital data capture, analysis, and integration allow for more efficient NDT processes, improved defect detection, and reduced inspection time.

Digitisation facilitates the collection and analysis of quality-related data, including inspection results, defect rates, and customer feedback. Advanced analytics techniques can be applied to identify patterns, trends, and root causes of quality issues. By analyzing historical data, manufacturers can gain insights into quality trends, identify areas for improvement, and implement corrective actions to enhance product quality.

Digitisation enables the implementation of statistical process control techniques for quality control. Real-time data from sensors and production systems can be analysed using statistical methods to monitor process variations, identify quality outliers, and trigger alarms when quality parameters deviate from acceptable ranges. SPC helps in maintaining process stability, reducing defects, and ensuring consistent product quality.

Digitisation allows for the seamless tracking and documentation of product information throughout the manufacturing process. Digital systems can capture data on raw materials, process parameters, inspection results, and certifications, ensuring traceability and compliance with quality standards. This information can be stored in a centralized database, enabling easy access for quality audits and customer inquiries. Digitisation facilitates supplier quality management processes. By integrating data from suppliers, manufacturers can assess and monitor supplier quality performance. Digital platforms can provide real-time visibility into supplier certifications, inspections, and quality metrics, allowing manufacturers to make informed decisions regarding supplier selection and collaboration.

Digitisation enables efficient collaboration and communication among stakeholders involved in quality control. Digital platforms and tools facilitate real-time sharing of quality-related information, inspection results, and corrective action plans. This improves collaboration between quality control teams, production departments, and suppliers, leading to faster resolution of quality issues. Digitisation promotes continuous improvement in quality control. By analyzing quality data and insights, manufacturers can identify areas for improvement, implement corrective actions, and monitor the effectiveness of quality control measures over time. Continuous improvement efforts driven by digitisation lead to enhanced product quality, reduced defects, and increased customer satisfaction.

By embracing digitisation and leveraging digital technologies, steel manufacturers can optimize their quality control and inspection processes. Real-time monitoring, automated inspection, data analytics, and collaboration tools enable proactive quality control, faster defect detection, and continuous improvement in product quality throughout the steel manufacturing process.

Collaboration and Integration

Digitisation promotes collaboration and integration across different stakeholders in the steel manufacturing ecosystem. This includes collaboration between equipment manufacturers, software providers, steel producers, and customers to develop standardized data exchange formats and interoperable systems.

Collaboration and integration are key components of the digitisation of steel manufacturing. By leveraging digital technologies and integrating various systems and stakeholders, manufacturers can improve communication, streamline processes, and enhance overall efficiency.

Digitisation enables the integration of the entire supply chain, including suppliers, manufacturers, distributors, and customers. By sharing data and information in real-time, manufacturers can have better visibility into the supply chain, streamline procurement processes, optimize inventory levels, and improve coordination with suppliers and customers. Digitisation involves the implementation of ERP systems that integrate different functions within the organisation. ERP systems provide a centralized platform for data storage and management, enabling seamless information flow between departments such as production, inventory, sales, finance, and human resources. This integration enhances collaboration, improves decision-making, and ensures data consistency across the organisation.

Digitisation enables collaborative planning and scheduling processes. Manufacturers can utilize digital platforms that allow different stakeholders, such as production managers, engineers, and supply chain teams, to collaborate on production plans, set priorities, allocate resources, and adjust schedules in real-time. This collaboration ensures alignment between production capabilities and customer demand, improving efficiency and reducing lead times. Digitisation facilitates the use of digital communication and collaboration tools, such as instant messaging, video conferencing, and project management platforms. These tools enable real-time communication, file sharing, and collaborative decision-making among team members, regardless of their geographical locations. Digital communication tools enhance information sharing, accelerate decision-making processes, and foster cross-functional collaboration.

Digitisation involves integrating production equipment with digital systems and sensors. This integration allows for real-time monitoring of equipment performance, predictive maintenance, and remote control. Manufacturers can collect data on machine health, utilization, and performance, enabling proactive maintenance, reducing downtime, and optimizing production efficiency. Digitisation involves the integration of sensors and IoT devices throughout the steel manufacturing processes. These devices can capture real-time data on parameters such as temperature, pressure, humidity, and vibration. The integration of sensor data with digital platforms enables data analysis, predictive analytics, and process optimisation.

Digitisation enables the integration of data from various sources, including production systems, sensors, and enterprise systems. By consolidating and analyzing this data using advanced analytics techniques, manufacturers can gain valuable insights, identify patterns, and make data-driven decisions. Data integration and analytics improve process visibility, quality control, and operational efficiency.

Digitisation encourages collaboration with research institutions, technology providers, and suppliers. Manufacturers can partner with research institutions to explore new technologies, conduct joint research projects, and stay updated on industry advancements. Collaboration with suppliers allows for the exchange of information, joint improvement initiatives, and better integration of the supply chain. Collaboration with research institutions and suppliers is an essential aspect of the digitisation of steel manufacturing. By partnering with research institutions and collaborating with suppliers, steel manufacturers can leverage external expertise, access cutting-edge technologies, and drive innovation in their digital transformation efforts.

Research institutions are at the forefront of technological advancements and innovation. By collaborating with research institutions, steel manufacturers can gain access to the latest research findings, emerging technologies, and industry best practices. This collaboration enables manufacturers to stay updated on the latest trends, explore new ideas, and adopt innovative solutions to enhance their digitisation efforts.

Collaborating with research institutions allows steel manufacturers to engage in joint research and development projects. These projects can focus on specific challenges or opportunities related to digitisation in steel manufacturing. By pooling resources, knowledge, and expertise, manufacturers and research institutions can work together to develop new technologies, improve processes, and address industry-specific issues.

Research institutions often develop novel technologies and solutions that can be valuable for steel manufacturers. Collaboration with research institutions facilitates technology transfer, where innovative solutions developed in the research environment are adapted and implemented in industrial settings. Manufacturers can benefit from the expertise of researchers and integrate cutting-edge technologies into their digitisation strategies.

Research institutions can provide testing and validation facilities to assess the performance and feasibility of new digital technologies in steel manufacturing. By partnering with research institutions, manufacturers can conduct pilot projects, test prototypes, and validate the effectiveness of digital solutions before large-scale implementation. This collaboration ensures that digitisation initiatives meet industry requirements and deliver the desired outcomes. Collaboration with research institutions fosters knowledge exchange between academia and industry.

Manufacturers can share their practical challenges and requirements, while researchers can provide insights, expertise, and training on digital technologies. This knowledge exchange helps manufacturers build internal capabilities, train their workforce, and develop a deeper understanding of the potential benefits and challenges associated with digitisation.

Collaboration with suppliers is crucial for successful digitisation in steel manufacturing. Suppliers play a significant role in providing equipment, software, and services that support digital transformation. By collaborating closely with suppliers, manufacturers can align their digital strategies, exchange information on requirements, and ensure seamless integration of digital systems and equipment into their manufacturing processes.

Collaboration with suppliers can go beyond traditional supplier-customer relationships. Manufacturers can engage suppliers in co-innovation activities, where both parties work together to develop customized solutions, optimize supply chain processes, and drive joint improvement initiatives. Co-innovation with suppliers leads to better integration, improved supply chain performance, and enhanced efficiency in the digitisation journey.

Suppliers often have access to emerging technologies and solutions related to digitisation. By collaborating with suppliers, manufacturers can gain early access to these technologies and assess their potential application in steel manufacturing. Early access to emerging technologies allows manufacturers to stay ahead of the curve, experiment with new solutions, and adopt innovative practices to enhance their digitisation efforts.

Collaboration with suppliers in the digitisation process enables manufacturers to optimize their supply chain operations. By integrating digital systems, sharing real-time data, and collaborating on inventory management, demand forecasting, and logistics, manufacturers and suppliers can streamline supply chain processes, reduce lead times, and improve overall efficiency.

Collaboration with research institutions and suppliers brings external expertise, research capabilities, innovation, and access to emerging technologies to the digitisation journey of steel manufacturers. By leveraging these collaborations, manufacturers can accelerate their digitisation efforts, drive innovation, and achieve greater success in transforming their operations.

Digitisation involves the creation of digital twins, which are virtual replicas of physical assets, processes, or systems. Digital twins enable simulation, modelling, and optimisation of manufacturing processes.

Manufacturers can use digital twins to test different scenarios, optimize production parameters, and improve process efficiency. By embracing collaboration and integration in the digitisation process, steel manufacturers can enhance communication, streamline processes, and drive operational efficiency. Integration of supply chain partners, systems, and stakeholders, along with the use of digital communication tools and data integration, leads to improved collaboration, faster decision-making, and increased productivity in steel manufacturing.

Advantages of digitisation

The digitisation of steel manufacturing offers numerous advantages that contribute to improved efficiency, productivity, quality, and sustainability. Digitisation streamlines processes, reduces manual intervention, and enhances automation in steel manufacturing. By integrating digital technologies, manufacturers can optimize production planning, scheduling, and resource allocation, resulting in improved efficiency, reduced downtime, and increased overall productivity.

Digitisation enables real-time monitoring and control of critical process parameters, leading to improved quality control in steel manufacturing. By utilizing sensors, data analytics, and AI-driven algorithms, manufacturers can detect deviations, identify defects, and take immediate corrective actions, ensuring consistent and high-quality output.

Digitisation allows for condition-based and predictive maintenance of equipment. By monitoring equipment health in real-time and analyzing data trends, manufacturers can predict maintenance requirements, schedule maintenance activities proactively, and minimize unplanned downtime. This leads to increased equipment reliability, reduced maintenance costs, and improved production efficiency. Digitisation generates vast amounts of data from various sources in steel manufacturing processes.

By leveraging data analytics and AI, manufacturers can extract valuable insights, identify patterns, and make data-driven decisions. This data-driven approach enhances operational efficiency, optimizes resource utilization, and supports strategic decision-making.

Digitisation facilitates seamless integration and coordination across the supply chain in steel manufacturing. By leveraging digital systems, manufacturers can enhance visibility, track materials, optimize inventory levels, and improve communication with suppliers and customers. This results in better supply chain management, reduced lead times, and enhanced customer satisfaction. Digitisation provides real-time visibility into production processes, equipment performance, and key metrics. Through IoT devices, sensors, and data analytics, manufacturers can monitor and control operations in real-time, allowing for proactive decision-making, immediate issue identification, and timely interventions to optimize production outcomes.

Digitisation offers opportunities for cost reduction in steel manufacturing. By optimizing processes, minimizing waste, and enhancing resource utilization, manufacturers can reduce production costs and increase profitability. Predictive maintenance and reduced downtime also contribute to cost savings by minimizing equipment failures and associated repair costs.

Digitisation supports sustainable practices in steel manufacturing. By optimizing energy consumption, reducing material waste, and implementing eco-friendly processes, manufacturers can minimize their environmental footprint. Digitisation enables the collection and analysis of data related to energy efficiency, emissions, and resource consumption, allowing for better environmental management and compliance with sustainability goals.

Digitisation enhances the agility and flexibility of steel manufacturing operations. Through digital systems, manufacturers can quickly adapt to changing market demands, modify production schedules, and respond to customer requirements in a timely manner. This agility allows manufacturers to stay competitive, address customization needs, and seize new business opportunities.

Digitisation fosters a culture of innovation and continuous improvement in steel manufacturing. By embracing digital technologies, manufacturers can experiment with new ideas, explore innovative processes, and drive continuous optimisation.

Collaboration with research institutions and suppliers further fuels innovation and supports the adoption of emerging technologies. The digitisation of steel manufacturing offers numerous advantages, including increased operational efficiency, enhanced quality control, improved decision-making, cost reduction, sustainability, agility, and a platform for innovation and continuous improvement. By embracing digitisation, steel manufacturers can transform their operations, gain a competitive edge, and thrive in a rapidly evolving industry landscape.

The digitisation of steel manufacturing is revolutionizing the industry, enabling steel producers to become more efficient, competitive, and sustainable. It enhances process visibility, control, and decision-making capabilities, leading to improved operational performance and customer satisfaction.

The digitisation of steel manufacturing has the potential to revolutionize the industry, improving productivity, efficiency, and sustainability while reducing costs and enhancing safety. It enables steel manufacturers to make data-driven decisions, optimize processes, and respond quickly to changing market demands. By embracing digitisation, steel manufacturers can stay competitive in a rapidly evolving global market.

The future of steel manufacturing

Innovation plays a crucial role in the digitisation of steel manufacturing. By embracing digital technologies and leveraging innovative solutions, steel manufacturers can transform their operations, improve efficiency, and drive growth. It involves the application of new ideas, technologies, and processes to transform and improve traditional manufacturing practices.

Digitisation enables the adoption of emerging technologies that drive innovation in steel manufacturing. Technologies such as the Internet of Things (IoT), Artificial Intelligence (AI), Big Data analytics, robotics, and automation offer new possibilities for process optimisation, data-driven decision-making, and enhanced productivity. By embracing these technologies, manufacturers can innovate their production processes, improve efficiency, and stay ahead of the competition. By integrating these technologies into their processes, manufacturers can optimize operations, enhance productivity, and unlock new opportunities for innovation.

Digitisation itself is a form of innovation in steel manufacturing. It involves the integration of digital technologies across various aspects of the manufacturing process, including design, production, logistics, and quality control. By digitally transforming their operations, manufacturers can streamline processes, improve data visibility, and unlock new opportunities for innovation and improvement.

Digitisation opens up avenues for product and process innovation in steel manufacturing. Digitisation opens up avenues for developing and implementing digital solutions tailored to the specific needs of steel manufacturing. Manufacturers can collaborate with technology providers and develop innovative software, platforms, and applications that address industry challenges. These solutions can range from real-time monitoring systems and predictive analytics tools to virtual reality (VR) simulations and digital twin technologies.

Manufacturers can leverage digital technologies to develop new products, enhance existing products, and optimize manufacturing processes. For example, digital simulation tools and virtual prototyping can facilitate the design and testing of new steel products, while process optimisation through data analytics and AI can improve efficiency, reduce waste, and enhance overall quality.

Digitisation encourages collaborative innovation by fostering partnerships with research institutions, technology providers, and other stakeholders. Collaborative innovation involves sharing knowledge, expertise, and resources to jointly develop and implement innovative solutions. Manufacturers can collaborate with external partners to explore new technologies, conduct research projects, and co-create solutions that address specific challenges in steel manufacturing. Digitisation enables process optimisation and continuous improvement in steel manufacturing. By leveraging data analytics, manufacturers can gain insights into process inefficiencies, identify bottlenecks, and make data-driven decisions to optimize operations. Innovations such as digital twins, simulation models, and optimisation algorithms help in analyzing various scenarios, improving process parameters, and driving continuous improvement initiatives.

Digitisation facilitates the implementation of smart and connected manufacturing concepts in the steel industry. Through the integration of sensors, IoT devices, and data analytics platforms, manufacturers can create a connected ecosystem where machines, equipment, and systems communicate and share data in real-time. This connectivity enables innovation in areas such as predictive maintenance, real-time monitoring, and adaptive control systems.

Digitisation fosters collaboration and co-innovation between steel manufacturers, research institutions, technology providers, and other industry stakeholders. By partnering with external entities, manufacturers can leverage expertise, access cutting-edge research, and collaborate on joint projects. This collaboration leads to the exchange of ideas, knowledge, and innovative solutions that drive the digitisation and innovation agenda in steel manufacturing.

Digitisation enables steel manufacturers to develop customer-centric solutions that meet specific market needs. Through data analytics and AI-driven customer insights, manufacturers can understand customer preferences, tailor products and services, and deliver personalized experiences. This customer-centric approach drives innovation by offering unique solutions, addressing customer pain points, and staying ahead of the competition.

Digitisation encourages open innovation and the development of ecosystems that bring together various stakeholders. Steel manufacturers can engage with startups, technology partners, research institutions, and industry associations to foster innovation. Open innovation platforms, hackathons, and innovation challenges provide opportunities for collaboration, idea generation, and the co-creation of innovative solutions.

Digitisation enables agile development and rapid prototyping of new ideas and solutions in steel manufacturing. Digital tools and platforms facilitate iterative development, allowing manufacturers to quickly test and refine concepts before full-scale implementation. This iterative approach to innovation accelerates the deployment of new technologies, processes, and business models. By embracing innovation through digitisation, steel manufacturers can unlock new opportunities, drive operational excellence, and create a competitive advantage.

Innovation in digitisation allows for the development of new products, improved processes, enhanced customer experiences, and sustainable growth in the steel manufacturing industry. Digitisation facilitates a culture of continuous improvement in steel manufacturing. Through real-time data monitoring, analysis, and feedback loops, manufacturers can identify areas for improvement and implement iterative changes. By embracing digital tools and analytics, manufacturers can gather insights, measure performance, and drive continuous improvement initiatives to optimize processes, enhance productivity, and achieve higher quality standards.

Digitisation enables greater customization and personalisation of steel products. By integrating digital design tools, manufacturers can offer tailored solutions to meet specific customer requirements. This customization capability fosters innovation by addressing unique customer needs and creating value-added products. Digitisation generates vast amounts of data throughout the manufacturing process. By applying advanced analytics and AI techniques to this data, manufacturers can gain insights, identify patterns, and uncover opportunities for innovation. Data-driven innovation allows for evidence-based decision-making, optimized processes, and the development of new solutions based on actionable insights derived from data analysis.

Digitisation enhances the agility of steel manufacturing operations, enabling manufacturers to adapt quickly to changing market demands and customer requirements. By leveraging digital technologies, manufacturers can respond rapidly to market trends, introduce new product variants, and adjust production schedules. This agility fosters innovation by facilitating quick experimentation, validation of ideas, and rapid adaptation to evolving industry dynamics. In summary, innovation is a key driver of the digitisation of steel manufacturing. By embracing emerging technologies, digital transformation, collaborative partnerships, and a culture of continuous improvement, manufacturers can innovate their processes, products, and business models to achieve higher efficiency, improved quality, and a

competitive advantage in the industry.

www.ingramcontent.com/pod-product-compliance
Lightning Source LLC
Chambersburg PA
CBHW082114220526
45472CB00009B/2178